AI-Powered Drones
Applications and Challenges

Table of Contents

Chapter 1. Introduction

In our rapidly progressing digital world, artificial intelligence (AI) continues to create transformative opportunities – a crucial one being in the world of unmanned aircraft systems, or drones. Our Special Report on "AI-Powered Drones: Applications and Challenges" delves into how AI melds with drone technology, catapulting its potential to new heights. This comprehensive exploration addresses everything from innovative uses in varying industries to the critical challenges faced in this integration, providing illuminating insights for both experts and aspirants in the field. To those grappling with the complexities of AI and drone technology, fret not, this report is designed to distill intricate concepts into an easily graspable narrative, inviting you on a riveting journey into one of the most exciting intersections of technology today. Buckle up and prepare to be amazed by how this captivating blend of AI and drone technology is reshaping our world, unveiling surprises at every turn, and continually pushing the boundaries of what we once thought possible.

Chapter 2. Introduction to Artificial Intelligence and Drones

Today, artificial intelligence (AI) and unmanned aircraft systems, better known as drones, represent two highly potent areas of technical innovation, each with its potential for transformative application. A close elucidation of these technology spheres provides us not only with an understanding of their intrinsic functionality but also unfurls a vivid panorama of their synergistic potential when melded together.

2.1. Understanding Artificial Intelligence

Artificial Intelligence, at its core, involves the creation of machines or systems capable of performing tasks requiring human-like intelligence. Such tasks may range from interpreting natural language and recognizing patterns to decision-making and problem-solving, amongst many others.

The basis of AI is machine learning (ML), a subset that enables machines to learn and improve from experience autonomically. AI systems are fed vast amounts of data, which they process to identify patterns, relationships, and underlying principles, thereby gaining knowledge and enhancing their problem-solving abilities. Deep learning, a further subset of machine learning, employs multilayered neural networks to mimic the human brain's functioning, recognizing even more complex patterns and making highly accurate predictions.

The journey of AI, which began as an exploratory concept in the mid-

20th century, has today escalated into a tangible reality that permeates every avenue of human life. From voice assistants like Alexa and Siri to fraud detection in banking transactions, artificial intelligence is showering its influence across diverse sectors, drastically altering societal norms and lifestyles.

2.2. Unveiling the World of Drones

When it comes to drones or unmanned aerial vehicles (UAVs), one tends to picture a small aircraft buzzing in the sky, typically employed for aerial photography or video capturing. However, drones embody much more than just airborne devices with imaging capabilities. Originating in the military domain for surveillance and combat missions, drone technology has over time evolved and perfected itself, infiltrating myriad civilian sectors.

Drones are capable of executing an array of operations that risk human lives, demand excessive time, or necessitate access to remote or unreachable terrains. From delivering packages and aiding in search and rescue missions to monitoring wildlife and providing aerial data for precision agriculture, drones are set to usher in a new era of technological advancements and possibilities.

2.3. The Confluence of AI and Drones

The coupling of AI and drone technology signifies a profound progression, setting the stage for unprecedented innovation and application. Where drones offer the physical mechanisms for varied operations, AI brings in the requisite intelligence, enabling these machines to perform tasks not just autonomously but also adaptively, enhancing their overall efficacy.

The union of these two spheres is geared towards more than just augmenting the autonomous capabilities of drones. AI-infused drones can recognize patterns and anomalies, undertake real-time data analysis, and make instant decisions, effectively carrying out tasks ranging from simple to complex.

For instance, AI-powered drones can detect unhealthy crops in a vast farmland through pattern recognition, pinpointing treatment areas. In logistics, they can adapt their routes in real time based on learned algorithms, effectively dodging obstacles or adverse weather conditions. The possibilities are as vast as they are varied, with each use case conferring substantial efficiencies and benefits.

□□□

Nonetheless, this transformative integration of AI and drones is not without its challenges, perceptible in the realms of ethics, laws, and technology. The subsequent sections of this report delve into the intricacies of these applications and challenges, elaborating on their mechanisms, implications, and future directions.

The engagement of AI and drones is a thrilling testament to human ingenuity and innovation, a fascinating precursor to an exciting era of technological evolution. As we journey into this new dawn of combined AI and drone capabilities, we can look forward to a world where the boundaries of what was once deemed possible keep getting shattered. This expedition would not just help us to understand these individual technologies better but also reveal the startling wonders of their confluence – a dazzling spectacle of advancement at its best.

Chapter 3. The Anatomy of AI-Powered Drones

AI-powered drones, fundamentally, are regular drones that have been augmented with artificial intelligence to permit autonomous decision-making and functionalities. The core concept of an AI-powered drone is the integration of AI routines, including Machine Learning, Deep Learning, and Computer Vision, into its operational systems.

3.1. Key Components of AI-Powered Drones

On the hardware front, AI-powered drones consist of eight crucial components: the chassis, propellers, motors, flight controller, battery, payload, transmitters, and receivers. The integration of AI is made possible by the incomparable fusion of this hardware with appropriate software.

Understanding each component:

1. Chassis: It provides the structure to which all other parts are attached and plays a significant role in the drone's stability and durability.

2. Propellers and Motors: Convert electrical energy into mechanical energy, generating lift.

3. Flight Controller: This is the command center of the drone, actuating the motors and coordinating with the GPS system to govern flight.

4. Battery: Powers all other components. With AI, optimal battery usage is a significant focus, as it enhances flight time and efficiency.

5. Payload: Represents the equipment drones carry based on their application. In AI-powered drones, payloads often include sensors and cameras for data acquisition and processing.

6. Transmitters and Receivers: These determine the range and reliability of control signals, while autonomous functions reduce the need for manual control.

Incorporated within the flight controller is the drone's software system, which includes flight control software, communication software, application-specific software, and AI software stack.

3.2. Software Systems in AI-Powered Drones

While hardware establishes the foundations of drone operation, it's the software that adds the intelligence part in AI-powered drones, enabling them to self-navigate, make autonomous decisions, and analyze information in real-time.

1. Flight Control Software: Controls the basic flight parameters, such as angle and speed. In AI-powered drones, it also loops in AI algorithms to determine flight path and allow for autonomous navigation.

2. Communication Software: Handles communication and data transmission between the drone and the base or among drones in a fleet.

3. Application-Specific Software: Tailored to specific use-cases such as mapping, agricultural analysis, infrastructure inspection, etc.

4. AI Software Stack: Houses a suite of AI algorithms that enable autonomous operations, image recognition, and real-time data processing.

The AI software stack, in particular, interfaces with other systems to equip the drone with incredible autonomy. Such a stack might incorporate Machine Learning, Deep Learning, and Computer Vision.

3.3. Machine Learning, Deep Learning, and Computer Vision in Drones

Machine Learning equips drones with the ability to learn from past experiences and improve performance without being explicitly programmed. When applied to drones, ML algorithms function as key enablers for autonomous navigation, pattern recognition, and decision-making.

Deep Learning, a subset of machine learning, uses artificial neural networks to simulate human decision-making processes. In AI-powered drones, DL algorithms play pivotal roles in image recognition and obstacle avoidance.

Computer Vision allows drones to interpret their surroundings by processing visual data. In conjunction with machine learning and deep learning, it enables advanced functionalities such as scene recognition, object identification, and even gesture recognition.

3.4. Challenges in Implementing AI in Drones

While AI integration has immense potential, it's not without challenges. AI algorithms demand high computation power, increasing the operational requirements of the flight controller and weighing the issue of power consumption.

Privacy and security also pose significant concerns. As these drones collect and store vast quantities of sensitive data, ensuring this data remains secure and is used responsibly is paramount.

Regulatory challenges also exist in many countries, with the laws often lagging behind the rapid advancements made by technology. Constructive dialogue must be established between regulators and innovators to facilitate growth in this sector responsibly.

The development and regulation of AI-powered drones represent a journey with subsequent challenges and rewards. As the field continues to evolve, we get closer to realizing the full potential of these remarkable devices. The road ahead is one of discovery, innovation, and boundless opportunity. The anatomy of AI-powered drones thus represents the intersection of intelligent software and sophisticated hardware that empowers our world, shaping our future towards new frontiers of possibility.

Chapter 4. Revolutionary Applications: From Agriculture to Zoning

The transformative potential of drones, when combined with the advanced capabilities of AI, has resulted in a myriad of revolutionary applications across various fields. This chapter delves into a detailed exploration of these, from agriculture to zoning.

4.1. Agriculture

Agriculture is at the forefront of AI-powered drone adoption. Precision farming techniques have been a sought-after tool by farmers worldwide, striving to optimize yields and methodically use resources. From surveying crops to monitoring soil health, AI-powered drones are proving to be indispensable in the agricultural industry.

Drones equipped with AI are able to detect patterns and make predictions, distinguishing healthy crops from those that are ill. This is possible through detailed image analysis using AI algorithms. By acquiring data from their surroundings, they provide farmers with a comprehensive birds-eye-view of their crop health, allowing for early interventions in case of crop diseases. This facilitates timely corrective measures that can save an entire harvest from being lost, proving invaluable to farmers.

Moreover, these drones go beyond visual survey. AI algorithms can analyze soil composition, enabling informed decisions about irrigations and fertilizers. This not only contributes towards the sustainability of farming methods by reducing wastage, but also ensures crops are getting the right balance of nutrients for optimal growth.

4.2. Building and Construction

AI-powered drones are altering the construction industry as we know it. From creating quick and accurate 3D maps, to surveying building site progress, to inspecting structures, they have quickly become an essential tool for contractors and architects alike.

With machine learning techniques, drones can now recognize patterns and discrepancies in building materials or constructions. Identifying such abnormalities at early stages can prevent construction errors, hazardous incidents, cost overruns, and delays. Furthermore, drones can provide real-time progress updates, ensuring construction projects stay on track.

4.3. Conservation

Conservationists are harnessing the power of AI drones to study and protect the environment in new, non-invasive ways. From tracking wildlife populations to identifying illegal activities, AI drones offer a novel approach to conservation activities.

Machine learning algorithms allow these drones to detect, count, and track species even in dense environments. Likewise, AI can differentiate between human activities, aiding in the identification and prevention of illegal activities like poaching or deforestation. This contributes massively towards biodiversity preservation efforts, expanding our capacity to monitor and guard the world's most precious ecosystems.

4.4. Disaster Management

Natural disasters are unpredictable and often devastating. Having robust systems for early warning, rescue missions, and damage assessment can mitigate the effects of such occurrences. AI-powered drones offer new possibilities in these aspects.

Equipped with thermal, infrared, and multispectral sensors, AI drones are able to cover large areas quickly, providing valuable data during times of crisis. AI further helps in interpreting this data rapidly, identifying areas that have been most affected, and aiding rescue teams in their coordination efforts. In the aftermath, the same drones can aid in evaluating the extent of the devastation, guiding reconstruction and recovery.

4.5. Entertainment

In the entertainment industry, drones equipped with AI have ignited a new era of creativity. From filming stunning aerial shots to creating synchronized drone light shows, AI technology is now an intricate part of the showbiz world.

AI helps these drones in optimizing flight paths, maintaining accurate camera angles, and assuring smooth, stable footages, required for professional-grade film productions. For light shows, AI-driven software ensures each drone is in the right place at the right time, creating breathtaking coordinated visual spectacles in the sky.

4.6. Zoning

Urban planning has long been a task of great complexity. AI drones streamline this process. They are capable of taking high-resolution aerial shots, enabling planners to view detailed images of city landscapes, which can then be used to inform zoning regulations.

Machine learning can extrapolate patterns from existing zones to predict optimal allocations for future city expansions, making urban planning far more efficient and less prone to human error.

In conclusion, the integration of AI with drone technology heralds new revolutionary applications in diverse fields of our day-to-day life. However, as we adopt such technologies, it's essential to navigate the challenges that come with them, ensuring that we leverage these tools responsibly and in ways that contribute towards the betterment of society.

The marriage of AI and drones promises to change the world as we know it, and we're only just scratching the surface of what's possible. Stay tuned for the journey ahead.

Chapter 5. Embarking on the Journey: Building an AI-Driven Drone

Bracing for an expedition in blending cutting-edge technologies involves a thorough understanding of each element's individual scope, functionality, and operation. An AI-driven drone is an ingenious confluence of artificial intelligence's cognitive intelligence and the mechanical versatility of unmanned aircraft systems.

5.1. Exploring the Core Components

Let's initiate to our journey by exploring the individual components of this blend. The fundamental building blocks of an AI-driven drone are made up of various intricate elements.

1. **Drone Hardware**: This is the physical body of the machine. Replete with motors, cameras, sensors, and mechanical controllers, the drone hardware is tailor-made for robust performance and efficient maneuverability.

2. **Embedded Systems**: These control the drone's movement, providing instructions on speed, direction, and altitude. They can also handle tasks like image processing, data collection, and communication.

3. **AI Modules**: The brain of the drone, the AI module, is responsible for the drone's intelligence. It's the command center, the place where decisions are made. Here, algorithms are devised to enable functionality from obstacle avoidance to autonomous navigation and beyond.

4. **Communication Tech**: This allows the drone to interact with a base station or operator. It's crucial for remote control, data

transmission, and in-dash control of autonomous functions.

5.2. Unpacking Drone Hardware

While the drone's anatomy may vary based on its application, certain key components remain constant. The propellers, motors, power supply, flight controller, sensors, and payload (e.g., a camera or other data collecting tools) are all vital parts of a standard drone setup. By thoroughly understanding the operating principles and functionalities of these elements, we are better equipped to conceptualize and design an intelligent unmanned aircraft system.

5.3. Learning the Language of Embedded Systems

In the heart of every drone's operation is the embedded system - the keystone to not just the drone's efficacy, but also its overarching autonomy. This system takes in sensor data, processes it, executes commands, and controls the drone's activities. Its primary functions include:

1. **Signal Processing**: Converting sensor data into useful insights that can be acted upon.

2. **Control System Design**: A set of rules guiding the drone's behavior based on the processed data.

3. **Controller Implementation**: Fine-tuning and setting the control algorithms into action.

Understanding and mastering embedded system design and operation is a crucial step to building an AI-driven drone.

5.4. Breathing Intelligence: AI Modules

Artificial intelligence in drones unravels a whole new territory of capabilities. With machine learning at its helm, drones are enabled to learn from their environment, improve their performance, and make informed decisions independently. The AI modules being integrated into drones today are based on neural networks that allow the drone to process visual data and navigate autonomously, perform object tracking and recognition, swarm coordination, and many more sophisticated operations.

5.5. Communication Tech: A Vital Link

Now let's turn our attention to the drone communication system. These high-frequency systems facilitate the drone's interaction with the operator panel and various other systems. Remote control, first-person view, and telemetry are some of the functions that heavily rely on drone communication tech.

5.6. Putting them Together: Designing an AI-Driven Drone

Once we have a firm grasp of these individual components, it's time to bring them all together. To architecture an AI-driven drone, we need to carefully lay out a plan that thoroughly discusses the purpose of the drone, the functions it needs to accomplish, and the tools it requires.

1. Start by defining the drone's purpose or mission. This will guide the type of hardware needed, the AI capabilities required, and the power and communication necessities.

2. Choose your hardware carefully. Remember, the drone's capability depends heavily on the equipment it's fitted with.

3. Design or select the correct AI modules. These need to be specialized and oriented to cater to the duties that the drone is expected to undertake.

4. Your communication system needs to be reliable, quick, and flexible enough to adapt to various scenarios.

5. Finally, test meticulously. Rigorous trials will ensure that your AI-driven drone can face real-world scenarios and respond appropriately.

The pathway to constructing an AI-driven drone is indeed complex, but by breaking it down into digestible and manageable pieces, the task becomes an enlivening challenge. Through this journey, we're pushing boundaries and transforming our technological landscape - so let's propel forward, with our drones leading the sky-bound charge.

Chapter 6. Data Collection and Processing: The Role of AI

Artificial intelligence (AI) plays a powerhouse role in every facet of drone operations, considerably enhancing their capabilities. However, one area where AI has an undeniable dominance is data collection and processing. Drones collect vast volumes of data during every flight from various sources like cameras, infrared sensors, and Lidar. Handling these streams of disparate data manually is a Herculean task. This is where AI steps in, not only streamlining the process of data collation but also turning it into actionable intelligence.

6.1. Harnessing AI in Data Collection

AI augments the data collection process in several ways. First, it helps in planning the drone's flight. Using sophisticated AI algorithms, an optimized path can be calculated that maximizes data collection while minimizing energy usage. This is essential in missions like crop inspection or geological surveying where large areas need to be covered. The drone can adjust its flight path automatically based on wind conditions, ensuring an effective data capture process. AI also enables drones to avoid obstacles during flight, which greatly reduces the risk of damage, and, by extension, loss of valuable collected data.

Another area where AI bolsters data collection is through sensor fusion - a process that amalgamates data from multiple sensors to provide more comprehensive outputs. For instance, an AI-powered drone might employ a combination of visual, infrared, and ultrasonic sensors. The AI system can analyze this varied data in conjunction, offering results that are far more nuanced than if the sensors were used separately.

6.2. Data Processing: AI-driven Transformations

Once data is amassed, it needs to be processed to extract meaningful insights. This stage is where AI's prowess truly shines. Advanced AI algorithms can sift through terabytes of data in minutes, a process that would take humans weeks or even months. Furthermore, the AI can conduct deep analyses, detecting patterns, correlations, and anomalies that are simply beyond human cognition.

Machine learning, a subset of AI, plays a crucial role in data processing. Through training, machine learning algorithms can learn to classify objects, detect features, and even predict future trends. Neural networks, a kind of machine learning, are particularly useful due to their ability to 'learn' complex relationships in data, akin to the human brain.

Consider image analysis as an application. Drones often capture thousands of high-resolution images during a survey. These images must be analyzed to uncover the information they hold. Traditional manual analyses are laborious and often unreliable due to human error.

AI dramatically enhances this process. With enough training data, an AI could learn to identify specific features in these images, whether it's differentiating between types of vegetation in an agricultural survey, detecting hotspots in a forest fire, or identifying signs of infrastructure wear and tear in urban planning. What's more, the AI can carry out these analyses in real-time during drone flight, allowing for immediate action or alterations in the mission as required.

6.3. Dealing with Unstructured Data

Much of the data collected by drones is unstructured. It comes in various forms like images, videos, thermal readings and does not adhere to a predetermined format. This poses a challenge; however, AI, specifically deep learning, is exceptionally adept at handling unstructured data.

Deep learning algorithms can learn to make sense of this raw, unstructured data, and classify it into useful categories. They can recognize features in images, understand changes in sensor readings, and make predictions based on patterns in data. These capabilities vastly simplify unstructured data handling, making AI pivotal in drone-based data acquisition and processing.

6.4. Tackling Limitations and Challenges

While AI significantly improves data collection and processing, there are still challenges. One is the computational demands of AI. Processing enormous datasets in real-time requires substantial computational power. Currently, most drones lack the on-board processing power necessary for complex AI operations, necessitating the use of remote servers.

Additionally, drone data quality is significantly impacted by environmental factors like weather, lighting, and turbulence. AI algorithms need to be robust enough to handle these variations. Achieving this robustness requires enormous volumes of training data, covering every conceivable scenario.

To conclude, AI's contribution to drone technology, specifically in data collection and processing, is monumental. It transforms drones from mere data collection tools to sophisticated analysis platforms. As AI technology becomes more advanced and drones more capable, the symbiosis between these technologies will undoubtedly continue to deepen, revolutionizing countless domains in the process. Despite the challenges in the path, the future of AI-powered drones in data acquisition and processing is luminous.

Chapter 7. Stellar Examples: AI-Powered Drone Success Stories

AI-powered drones have been instrumental in various sectors globally, continually affirming their value with undeniable success stories. These instances of achievement not only highlight their potential functionality but also teases the boundless possibilities lying ahead, waiting to be untapped.

7.1. Precision Agriculture and Environmental Management

In the realm of agriculture and environmental management, AI-powered drones have been transformative. One stellar instance is the case of a Californian vineyard where the typical manual methods of examining the vast plantation were replaced with drones equipped with advanced sensors and AI algorithms. These drones mapped the plantation while capturing high-resolution images. The AI, trained in image recognition and data analysis, could process these images to identify plant diseases, variations in plant water need, and even predict the optimal time for harvesting. This incorporated methodology improved crop yield while reducing the amount of water and pesticides used, profoundly transforming the vineyard's productivity and sustainability.

7.2. Drone Delivery Service

An exciting success story in the field of AI-powered drones delivery is from the renowned company, Zipline in Rwanda. Established in 2016, Zipline has managed to create the world's most comprehensive drone delivery service, not for food or consumer products, but for life-saving medicines. Using AI and drone technology, Zipline delivers vital medical supplies (such as blood transfusions, vaccines, and medications) within minutes, across difficult terrains and inaccessible regions, thus saving numerous lives. It exemplifies the vast potential of AI and drone technology in delivering critical services.

7.3. Infrastructure Inspections

AI-powered drones have remarkably improved infrastructure inspection practices. A case in point is the application of AI-enabled drones in inspecting wind turbines in Scotland. These drones equipped with AI and image processing abilities, analyze high-definition images of the turbines to detect any faults or irregularities efficiently and safely. Not only has this drastically cut the time for inspection, but it also ensures the safety of the inspection personnel. This drone application projects the potential AI-enabled drones have in reducing risk and augmenting efficiency in infrastructural management.

7.4. Disaster Management and Relief Services

In the domain of disaster management, the use of AI-powered drones has been remarkably effective. In the aftermath of the devastating earthquake in Nepal in 2015, teams deployed AI-powered drones that could map the area quickly, identify impassable routes, and locate survivors under rubble using AI-powered image analysis techniques. This quick and effective response greatly mitigated the loss of life and facilitated rescue efforts. This instance highlights the potential drones and AI have in improving disaster response strategies.

7.5. Wildlife Conservation

AI-powered drones' success story would be incomplete without mentioning their contribution to wildlife conservation. In Africa, drones fitted with AI-powered thermal imaging cameras are used to track night-time movements of endangered species and detect poachers. The AI processes the thermal images, differentiating between animal species and humans based on their thermal signatures. The drones also act as deterrents to poaching activities, thus playing an active role in wildlife protection and preservation.

In the grand scheme of technology, AI-powered drones are still relatively new, and with these diverse success stories, we have just scratched the surface of what's possible. In a world often daunted by the spectre of AI, these drones have shown their potential to be instruments of change, optimisation and protection. Each success paves the way for us to envision and create a world where AI-powered drones are a key player in solving complex problems and increasing global accessibility, efficiency and safety.

This is a testament to the promises AI holds in the practical world. From growing our food with precise care, ensuring that medical supplies reach the remote corners in time, improving the efficiency of our infrastructures, being an aid in the face of disasters to protecting our wildlife, AI-powered drones are steadily becoming a staple tool of a progressive society. With constant advancements and exploration in AI and drone technologies, these tales of success are just the opening chapter in a long, exciting saga of innovation waiting to unfold.

Chapter 8. Navigating Through Airspace: Legal and Regulatory Considerations

Unmanned aircraft systems (UAS), commonly called drones, operating in conjunction with artificial intelligence (AI) offer a broad gamut of opportunities across various industries. Nevertheless, to facilitate safe, efficient, and legal use, it's paramount to understand the air laws and regulations guiding their use.

8.1. Air Law Framework

The understanding of air laws starts from the fundamentals of airspace classes. Different classes of airspace are established based upon the complexity, volume, and nature of aircraft operations within these classes – ranging from Class A, the most regulated, to Class G, the least. Drones, primarily operate in Class G airspace, unless approved by aviation authorities.

Air law, across many nations, falls under the arm of federal control. In the United States, the Federal Aviation Administration (FAA) is vested with the authority, while in countries like the UK, India, and Australia, it's the Civil Aviation Authority (CAA), Directorate General of Civil Aviation (DGCA), and Civil Aviation Safety Authority (CASA), respectively. These regulatory bodies implement policies, regulations, and protocols to ensure safety in the aviation domain.

Government authorities generally establish criteria like the drone's purpose (commercial, recreational, or public), operator certifications, weight of the drone, and the operating environment. Furthermore, each category engenders its unique set of rules and guidelines for operation.

8.2. Acquiring Operator Certification

For commercial drone operations, the operator must obtain relevant certifications or licenses. As an illustration, in the U.S, the FAA mandates operators to hold a Remote Pilot Certificate, also known as Part 107 certification. This certification requires applicants to pass an aeronautical knowledge test and verify their identity to be deemed qualified remote pilots.

Similar certification formats are applied in various nations, albeit sometimes under different names. These certifications reinforce that the operator possesses adequate knowledge about drone operations and complies with the respective aviation authority's guidelines.

8.3. Registration and Marking Requirements

Most regulatory authorities stipulate the registration of drones, especially those over a certain weight threshold. In the U.S, the FAA requires all drones weighing more than 0.55 pounds to be registered. Registrations help quickly identify owners in the event of an accident or misuse.

Post registration, the drone is to be visibly marked with the registration number. It assists the regulators and other parties to trace any drone involved in an incident back to its pilot.

8.4. Flight Permissions and Restrictions

In most regions, commercial drone operations necessitate flight permissions, particularly for areas outside designated zones, flights beyond visual line-of-sight (BVLOS), or flying high-risk operations. It's prudent to understand these specific permissions to avoid legal complications.

For instance, the FAA's LAANC (Low Altitude Authorization and Notification Capability) allows real-time processing of airspace authorizations below approved altitudes in controlled airspace, negating the need for a manual application process.

Also, there are geospatial restrictions— no-fly zones over critical infrastructure, populated areas, and sensitive government locations. These measures are there to eliminate risk factors and bolster public safety.

8.5. Use of Artificial Intelligence

AI comes into the regulatory limelight due to its scalability and decision-making capabilities. It brings about considerations like data privacy, ethical use, and legal liability in case an AI-controlled drone fails, causing damage. Algorithms used in AI are usually opaque and difficult to understand, causing regulation complications.

However, as AI evolves, regulatory bodies are gradually developing frameworks to address these issues. The focus is on ensuring transparent AI operations and maintaining the responsibility of human operators.

8.6. Liability and Insurance

Liability comes into play when a drone causes damage or harm. In such scenarios, insurance can be a valuable safeguard.

Although not universally mandatory, it's considered a best practice for commercial drone operators to maintain liability insurance. The cover usually extends to bodily injury, property damage, and privacy claims.

A well-rounded understanding of air laws and regulations, liability, and insurance, allows drone operators to exploit the potential of AI-powered drones while maintaining adherence to safety norms and regulations. Ethical, legal, and adept operations will be key to fostering trust and broadening the scope of this technology in the years to come.

Despite the complexities, these regulatory measures are necessary to ensure a safe drone ecosystem, protecting the operators, people on the ground, and other airspace users. The future looks exciting for AI-powered drone technology, and with every measure taken towards legal and safe operations, we journey one step further into that future. As technology keeps advancing, so will laws and regulations, reframing not just our skies, but also how we navigate our world.

Chapter 9. Challenges and Concerns: The Dark Side of AI Drones

As we continue to push the boundaries of technological capabilities with artificial intelligence (AI) and drone technology, it's essential to note that this exciting blend doesn't come without a share of challenges and concerns. While the potential applications indeed hold the power to revolutionize industries, so too do they present a plethora of new problems demanding our immediate attention and viable solutions.

9.1. Regulatory Hurdles

Central to the deployment of AI drones is compliance with regulatory guidelines and standards. Countries across the globe have varying levels of regulations related to drone usage. These laws dictate everything from who can operate a drone, under what conditions they can fly them, and even the type of data that can be collected. This unfixed regulatory landscape not only complicates uniform adoption but also presents regular compliance issues for users.

Regulations also grapple with the rapidly evolving technology itself. Current guidelines are often found lacking when faced with the new capabilities that AI imbues unmanned aircraft with, such as Beyond Visual Line Of Sight (BVLOS) flight, autonomous navigation, and real-time data processing.

9.2. Privacy Concerns

A critical concern arising with the proliferation of drones is the invasion of privacy. Drones equipped with AI-enhanced surveillance technology can gather comprehensive data, often discreetly and without the knowledge or explicit consent of the people involved. This issue escalates when drones are used to intrude upon private spaces, resulting in violation of privacy rights.

Furthermore, the information that these drones collect, given their wide range of sensory capabilities, isn't limited to images or videos. They can capture data concerning heat signatures, network signals, and even seismic activity. Such extensive data collection brings with it a heightened responsibility of usage and storage, with unauthorized access or misuse posing significant risks.

9.3. Ethical Questions

Another discursive area is the ethical questions that drone use generates. Military drones, for instance, have been a subject of controversy with regards to collateral damage, robotic warfare, and the psychological impact on drone operators.

Moreover, the incorporation of AI into drones hands them a high level of autonomy. In situations where these autonomous drones may need to make decisions that have ethical implications – such as avoiding a crash or a collision – the question naturally arises of whose ethics the machine is following, and whether these decisions are justifiable on a moral landscape.

9.4. Security Risks

The flip side of AI drones' efficiency is the vulnerability they introduce into our security systems. Cybercriminals can hack into these drones, taking over controls, manipulating data, or even using them as weapons. As these high-tech devices become more sophisticated, so do the strategies deployed by malicious actors.

Furthermore, drones could also be used for corporate and

governmental espionage. Their ability to infiltrate physical spaces without detection makes them ideal tools for such illicit purposes.

9.5. Technological Limitations

Finally, while AI-powered drones hold tremendous potential, the technology itself is not flawless. Machine learning algorithms are reliant on the quality and amount of data they're fed. They can produce inaccurate or biased results due to insufficient training or skewed datasets.

Simultaneously, AI-powered drones demand a high level of computational capability and battery life, areas where limitations can significantly hamper operations. As such, ongoing technological innovations will be key in overcoming these hurdles and unlocking the full potential of AI drones.

In conclusion, this brave new world of AI-power drones holds a captivating blend of potential and challenges. Our ability to harness the benefits while effectively navigating these issues will determine how far and how fast we can move ahead. This complex narrative continues to unfold, urging us to balance our technological ambitions with the ethical, security, privacy, legal, and humanitarian considerations that come with this formidable duo of AI and drone technology. The darkness lingers not because of the limitations of our technology but due to our grappling with the profound questions and challenges it presents. As we continue this exploration, we must do so with prudence, responsibility, and a deep respect for the transformative potential it holds.

Chapter 10. The Road Ahead: Predictions and Potential

A glimpse into the future of AI-powered drones exposes a world where limits continue to be pushed, and capabilities persistently grow. This chapter strives to explore the colossal potential affiliated with the blending of AI and drone technology, paving a path into unexplored avenues of usage and exploration, whilst acknowledging and providing solutions for the complex issues we are bound to encounter.

10.1. Potential

Defining the potential of AI and drone technology is akin to contemplating the endless universe: mind-boggling and full of surprises. Research and development are consistently broadening the horizons, with the technology being embraced in various sectors like agriculture, real estate, military, and ecommerce, to name just a few. Below is an exploration of a few key arenas where we can forecast the technology's vast influence.

10.1.1. Agriculture

The implementation of drone technology in agriculture has already proven revolutionary, with farmers using these unmanned aircraft for tasks such as crop inspection and planning. With AI integration, these applications are poised to become even more robust. Analyzing enormous tracts of land, drones could identify anomalies in crop growth, water shortages, or flawed irrigation systems with high precision. The potential lies in their ability to transform the farming operations from reactive to proactive, identifying potential issues even before they surface.

10.1.2. Transportation

Another sector forecasted to dramatically benefit from the AI and drone blend is transportation. Autonomous drone taxis and delivery drones will not only reshape urban mobility but also extend services to remote areas lacking robust transport infrastructure. In package delivery, drones fitted with AI can navigate autonomely and efficiently, reducing carbon emissions and drastically increasing the speed of delivery times. The future could even see drones facilitating critical services like medical supply transportation in hard-to-reach areas.

10.2. Predictions

The impending possibilities outlined above merely scratch the surface of the vast potential for AI-powered drones. Yet, pinpointing exact timelines for these advancements is challenging. Technological development rarely follows a linear trajectory, and unexpected breakthroughs could accelerate growth. However, a cautious yet optimistic forecast suggests some key developments.

10.2.1. Increased Autonomy

By 2030, it is plausible that significant advancements in AI technology will diminish the need for human control, and drones will operate with heightened levels of autonomy. Combining advances in machine learning, computer vision, sensor technology, and control systems, drones of the future will be capable of more complex operations, making critical decisions in real-time and navigating more hazardous environments autonomously.

10.2.2. Legislation and Regulation

Legislation trail the strides in technological advancement, and so must catch up. By the close of this decade, a more comprehensive legal framework for the uses of AI-powered drones is anticipated. These would address issues of privacy, safety, and noise pollution, among others. Anticipation is that regulations will become stringent and detailed, but also more accommodating of the wide range of drone applications.

10.3. Challenges and Solutions

The envisioned road ahead is no bed of roses. To actualize this high potential, several key challenges need addressing. Safety and privacy concerns rank among the most significant considerations, closely followed by noise pollution, and dependence on weather conditions. Solutions to these issues require technological innovation, calibrated legislation, and public acceptance shaped by transparent communication.

10.3.1. Safety

One of the paramount challenges is that of safety. The proliferation of drones opens up the risk of collisions, crashes, and general air traffic management issues. This is where AI can offer a solution: using complex algorithms, drones can be equipped to perform real-time navigation and collision avoidance. As for air traffic management, creating airspace corridors for drones, or a principle of 'sky lanes,' might be a valid solution.

As we stride into the uncharted territories of combining AI with drone technology, the road ahead brims with sweeping opportunities and equally challenging obstacles. Being at the forefront of this technological revolution, it is our responsibility to push for continuous innovation and responsible use, ensuring a future where AI-powered drones take us to new horizons while respecting the vibrant tapestry of human values and societal norms present globally. The road to achieving this will be long, and yes, arduous at times. However, the benefits, from improved farming practices to expanded transportation methods, far overpower the struggles, promising a future as fascinating as it is remarkably transformative.

Chapter 11. Conclusion: Impact of AI Drones on Future Society

Artificial Intelligence (AI) in tandem with drones is already precipitating significant changes in numerous industries. With the advent AI-powered drones, many of these advancements are steadily becoming embedded in the fabric of our societies, providing the impetus for a transformative shift from the norm. It is inescapable that the impact of AI drones on the future society will be profound, to say the least.

11.1. Analyzing the Economic Impact

The mainstream adoption of AI drones heralds significant economic implications. A PWC report suggests an estimated $127 billion worth of labor and business services could be replaced by drone technology. This isn't necessarily a dire circumstance as it appears – the economic value directly contributes to increased efficiency and effectiveness, leading to net positive outcomes.

AI drones can execute tasks faster and with reduced resource expenditure, thus acting as economic stimulants – fostering new investments, creating jobs, and spurring innovation. Pioneers in drone technology can leverage these economic opportunities, shaping resilient, technology-driven economies.

In agriculture, drones equipped with AI-powered imaging sensors help to monitor crops and livestock in real-time, surpassing traditional methods in efficiency. In logistics and supply chain, companies like Amazon are already conducting final-mile delivery trials with drones, signaling a paradigm shift in business operations. Such industry-specific applications are arguably driving a significant avenue of economic growth.

However, it is also important to tread this path with caution. The transition must be managed carefully to mitigate potential job losses in the short term and ensure that new job opportunities created by AI and drones are accessible to those displaced from traditional roles.

11.2. Societal Impact: Democratizing Information

AI drones are influentially democratizing information access. Advanced data capture and processing capabilities empower drones to collect and analyze data in real-time, eliminating time-consuming traditional processes and transforming decision-making strategies across industries.

For instance, in the world of journalism and broadcasting, drones are transforming the way we gather news. They provide unsurpassed access and perspective to events unfolding in real-time that might otherwise be difficult or dangerous to reach, delivering on the promise that up-to-date and accurate information is the right of every citizen.

11.3. Potential Challenges

Despite the promise of unprecedented social and economic growth, there also lies a multitude of challenges that humanity needs to address proactively.

Privacy is a major concern when it comes to drone usage. Unregulated drone flights could intrude into private spaces. Therefore, crafting robust privacy regulation which balances the benefits against the risks, is the need of the hour.

Security threats are another alarming challenge. In the wrong hands, AI drones could potentially be weaponized or utilized for illicit activities. While the opportunities are endless, the balance of power this technology can lead to must not be overlooked.

Moreover, not everyone is prepared for the drone revolution. The digital divide can exacerbate existing inequalities if access to drone technology is not made equitable and inclusive. Governments alongside private entities need to invest in digital literacy, ensuring that the effects of AI drones can touch the lives of all individuals positively.

11.4. Navigating the Future

The integration of AI with drones is ripe with potential, persistently pushing the boundaries of our societal norms - thereby cultivating a future interlaced with technological advancements. It conjures up a future where unmanned drones are commonplace: delivering packages, assisting in agriculture, aiding in disaster management, and even serving as alternative modes of transportation.

However, it is crucial to remember that the overall impact of AI drones on the society of the future will be shaped not only by technological advancements but also by how we adapt to and shape the use of this technology. This locus of development necessitates an open dialogue about the ethical implications, security concerns, regulatory standards, and potential for advancement.

In summing up, AI drones' impact on future society is potentially boundless. The path that lies in wait is navigated by a myriad of multifarious considerations, shaped by equal parts promise, potential, and prudent caution. As such, the final trajectory resides not just in the technology itself – but in the hands of those driving its adoption. It is a testament to human innovation and ingenuity, a tool wielded as much for its power to push technological boundaries as for its responsibility in guiding societal norms and values responsibly into the future.